The Planner HQ Guide To:

MANAGING THE WORK ORDER LIFECYCLE

Chris Diprose

A Planner HQ Publication

PLANNER HQ

TOOLS AND TRAINING FOR PLANNING EXCELLENCE

ISBN: 978-1-9192306-2-7

Published by: Planner HQ

For more resources, visit: www.theplannerhq.com

Follow us on LinkedIn:

About Planner HQ

Planner HQ is a professional development platform built to elevate the role of maintenance planners and those working at the heart of reliability, asset management, and operations.

The goal is simple: to empower maintenance planners to plan better, lead better, and deliver meaningful impact in their organisations. Through practical tools, in-depth training, and thought leadership grounded in real-world experience, Planner HQ helps individuals and teams move beyond reactive maintenance and into structured, proactive, and data-driven execution.

The project was created by people who have lived the planner role across industries, across continents, and across systems - from pharmaceutical plants and manufacturing lines to national healthcare systems and major transportation infrastructure. That real-world foundation is what makes Planner HQ different. It's not just theory - it's about what works.

Whether you are an individual planner looking to sharpen your skills, a leader seeking to upskill your team, or an organisation aiming to improve reliability outcomes, Planner HQ is designed to support your journey.

To learn more, explore future tools, and stay connected with the evolving Planner HQ ecosystem, visit us at:

→ www.ThePlannerHQ.com

Introduction

Why This Handbook Exists

If you're reading this, you already understand one thing: work orders drive everything.

As a maintenance planner, scheduler, or coordinator, you're the one responsible for turning vague requests into structured plans. You bridge the gap between what people say needs doing - and what actually gets done. And that journey, from request to completion is defined by your ability to manage the work order process with clarity and discipline.

A little too often, the work order lifecycle is messy, misunderstood, or mismanaged. Work requests come in with missing detail., priorities are unclear, jobs hit the schedule before they're truly ready. And Work Order close-out; the quality and detail just isn't where it should be if we are talking about running world class maintenance management processes.

This handbook exists in some part to fix that.

Managing the Work Order Lifecycle is about building structure into the chaos. It's about helping you bring order, flow, and accountability to every stage of the process - from the moment work is requested to the point it's closed, captured, and complete.

This isn't too heavy on theory, instead leaning towards practical strategies, simple visuals, and clear checkpoints to help you manage the work order lifecycle with confidence in your plant, with your team, under real-world pressure.

Who This Book Is For

This handbook is primarily written for maintenance planners - the people on the frontline of work control, job flow, and schedule readiness.

However, its relevance stretches further. Whether you are:

- A planner/scheduler managing the CMMS and weekly workload
- A supervisor balancing tools and team leadership
- A reliability coordinator looking to align planning and failure data
- A technician stepping into a hybrid planning role
- An operations or engineering lead who needs better alignment with maintenance

This book is for you. If you touch work orders (request them, plan them, schedule them, track them) you'll find tools here to make that process better, tighter, and more reliable.

A Chapter a Day, a Better Lifecycle by Week's End

This handbook follows a 7-chapter format - not just for structure, but for momentum.

You can tackle one chapter a day and come away in a single week with a clearer, stronger approach to the work order lifecycle. One chapter, one stage, one improvement at a time.

Whether you're reading straight through or jumping to the stage you're struggling with, the format is flexible, but the value stacks as you go - because every piece of the work order lifecycle connects to the next.

How to Use This Book

Each chapter tackles one part of the work order lifecycle - from capturing and triaging requests, through planning and execution, all the way to close-out and continuous improvement.

You'll find:

- Clear explanations and real-world scenarios
- Visual diagrams and process flows
- Practical guidance that's easy to apply

And at the end of every chapter, you'll find a Planner Exercise - a short, focused set of reflection prompts to help you apply the content in your own context. Use the lined notes page to jot down challenges, ideas, actions, or observations.

Revisit these regularly. These small moments of reflection often lead to your biggest breakthroughs.

You can use this handbook:

- At your desk while reviewing open work
- In team meetings as a reference
- During audits or process reviews to strengthen your approach
- Or as a tool to coach others into better habits

This is part of the *Planner HQ Guide To* series - a growing library of focused resources built for modern planners who want more control, less chaos, and better results.

If you're ready to bring discipline, structure, and clarity to the way work orders are handled in your organisation - this is where it starts.

And if you haven't yet explored *The Maintenance Planner's Playbook*, that's where you'll find the full planner role blueprint. Together, these tools form the backbone of Planner HQ's mission: to support maintenance planners with the tools, training, and structure they've been missing for too long.

Sign up to the Planner HQ email list to access free downloads, bonus tools, and future releases - and take your next step toward planner mastery.

Because better maintenance starts with better work orders. And better work orders start with you.

Table of Contents

Chapter 1

Understanding The Work Order Lifecycle

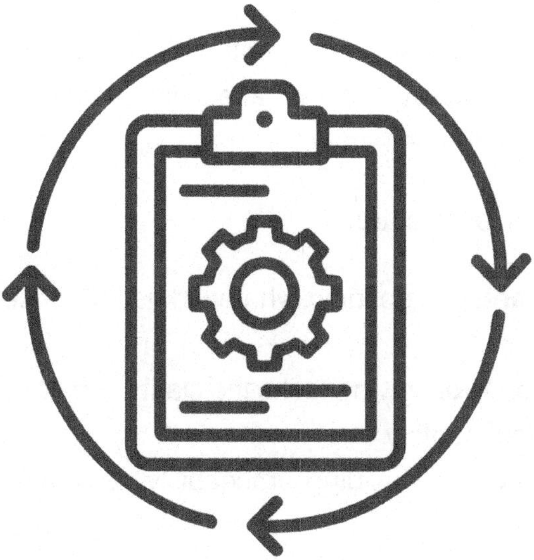

Work orders are the heartbeat of your planning system.

Whether you're mapping out a six-month PM schedule, reacting to an urgent breakdown, or slowly burning through a mountain of corrective work in the backlog, it usually all comes back to the same tool: the work order.

Every task that gets done (or doesn't get done) flows through a work order in some shape or form. It's the foundation of how we assign responsibility, document actions, and measure performance. Despite their critical role, work orders are one of the most misunderstood tools in the entire maintenance process.

Some see them as just admin, digital paperwork to be filled in because "that's the process." Others treat them as a one-size-fits-all form - something that exists because the CMMS needs something to schedule against. Unfortunately, that mindset really misses the bigger picture.

A work order isn't just a box to tick. It's the central thread that links together maintenance strategy, daily execution, resource planning, and continuous improvement.

When used properly, work orders are:

- **A control mechanism** - defining what work is authorised and why
- **A communication tool** - giving techs the clarity and scope they need to execute safely
- **A historical record** - building traceability and asset knowledge over time
- **A measurement tool** - turning work into data, and data into strategy

That's why planners must understand the full lifecycle of a work order, not just the middle stage where planning takes place. When you only focus on writing tasks and scheduling labour, you miss the upstream and downstream issues that can make or break your entire planning effort.

You can plan the best job in the world, but if the request was vague or unauthorised, or if the close-out data is skipped, the value of that plan disappears fast.

This chapter is about zooming out and understanding the work order end to end.

? What exactly are work orders?
? Why are they so important in a CMMS environment?
? How should planners engage with them. Not just reactively, but with ownership across the entire lifecycle?

Let's start by unpacking what work orders really are - and how they move from start to finish.

What Do They Look Like?

While layouts vary by CMMS, a typical work order will include:

- Work Order number
- Asset or equipment ID
- Requestor information
- Problem description
- Priority or work type
- Assigned technician or team/shift
- Planned date and duration
- Job plan or instructions
- Safety and permit notes

- Parts and materials
- Completion notes and close-out fields

The exact layout might differ depending on your organisations CMMS - but the function is universal. It's a shared language for maintenance teams, supervisors, operations, and auditors.

Types of Work Orders

Work orders don't all serve the same purpose. Your CMMS may support a wide range of types, but most fall into a few common categories:

- Proactive Maintenance
    - **System Checks (SC):** Non-intrusive running checks, often just visual.
    - **Preventive Maintenance (PM)**: Recurring work based on time, usage, or compliance.
    - **Condition-Based Maintenance (CBM)**: Advanced techniques using vibration, thermal etc.
- Reactive Maintenance
    - **Corrective Work**: Non-urgent fixes found during PM or inspections.
    - **Breakdowns / Emergency Repairs**: High-priority work due to failure or downtime.
    - **Unplanned Requests**: Triggered by operators, inspections, or site walkdowns.
- Project or Campaign Work
    - Multi-day, multi-scope work requiring coordination (e.g. equipment upgrades, shutdown prep)
    - Safety actions
    - Audit findings

Work order types will vary between CMMS platforms and industries, but the point remains the same: **not all work orders are equal** and understanding their purpose is essential to managing them well.

The Full Lifecycle of a Work Order

A work order isn't just something you raise and forget. It goes through a series of stages - each with different inputs, owners, and risks. Understanding this flow is the key to controlling your planning process.

Here's the general lifecycle (for a work order on a time/condition based frequency):

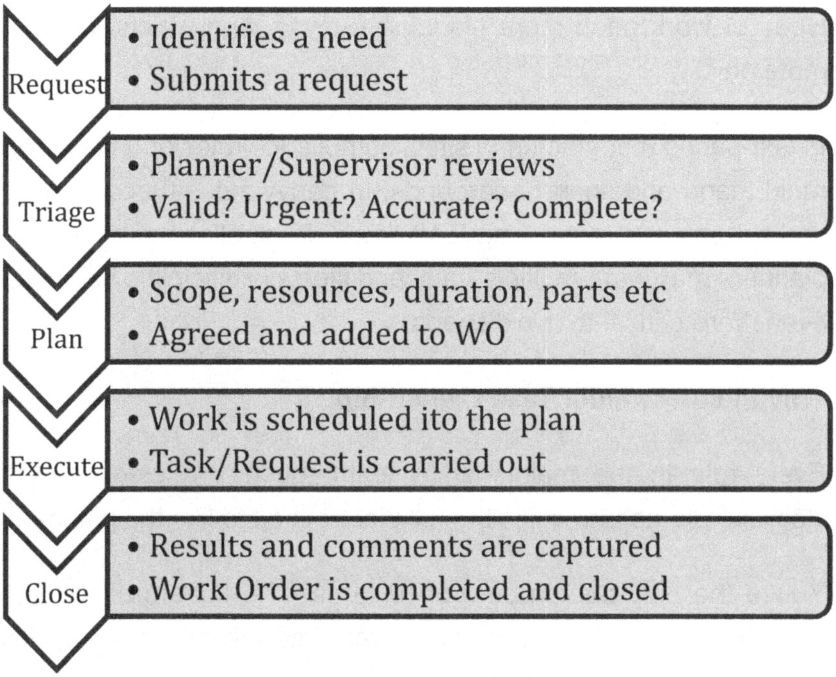

Each of these stages may vary in length and complexity depending on the site, team, and asset - but this structure applies in almost every environment.

How Long Does It Take?

There's no single right answer.

- **Corrective work** might take a week or two depending on parts or access.
- **Breakdowns** are usually immediate and short-lived.
- **Project work** could last months and be split across many WOs.

A planner's job is not just to track time, but to understand which types of work need more planning runway, and which can flow more quickly.

In fast-paced manufacturing sites, compliance rules often demand rapid triage and short turnarounds. In heavy industry, complexity and access can mean long lead times. In smaller organisations, planners may wear multiple hats, and lifecycle discipline becomes even more critical to avoid chaos.

Why Planners Must Take Ownership

Every role in the maintenance team touches a piece of the lifecycle, but no one owns the full picture it quite like the planner.

You're the only person positioned to see the full arc of the job. From the moment a request lands, you determine if it's valid, if it's ready, if it fits the schedule, and how it gets closed.

That means:

- Pushing back/rejecting unready or vague requests
- Ensuring proper planning before the schedule is locked
- Following up on close-out accuracy and data hygiene
- Driving improvements to reduce repeat issues or inefficiencies

It also means recognising when the process is broken and knowing how to fix it.

Why Planners? Because the planning role combines technical understanding, system fluency, operational awareness, and the strategic mindset to make this flow work. No one else brings that blend to the table.

That said, planner do not own this process alone.

- **Operations** must raise good, actionable requests
- **Supervisors** must help validate, prioritise, and execute
- **Technicians** must follow the plan and close work with quality data
- **Reliability Engineers** feed insights back into better planning

It's the planner who ties it all together. The lifecycle succeeds or fails based on how well the planner drives structure and clarity through it.

A Framework for Discipline and Reliability

Once you start viewing work orders as a lifecycle, not just a list, everything shifts.

Suddenly, you stop seeing work as a pile of tasks - and start seeing it as a flowing system. One where every stage affects the next, and where gaps at any point create ripple effects that derail your whole week.

You begin spotting the patterns where things fall apart:

- **Work is accepted before it's ready** - with missing info, unclear scope, or no priority
- **Jobs are scheduled with missing parts** - resulting in rework, delays, and lost credibility
- **Technicians receive vague instructions** - leading to improvisation, skipped steps, or safety risks
- **Close-out data is incomplete or skipped** - making analysis, audits, or PM reviews worthless

But once you frame the process as a lifecycle, you naturally start building discipline into each step. You stop chasing fires and start controlling the flow.

What that discipline looks like:

- **Standard triage**: Clear criteria for accepting or rejecting incoming work
- **Planning gates**: A defined checklist before work is eligible for scheduling
- **Status workflows**: Visual, trackable steps that show where each work order is in the process
- **Ownership and follow-through**: Everyone understands their role - and what's expected from them

Instead of relying on tribal knowledge or last-minute saves, you're establishing a repeatable pattern. A system that builds trust with technicians, supervisors, and stakeholders because it's consistent, visible, and built to last.

This structure also gives you leverage. You can scale your planning workload more effectively, onboard new team members faster, and identify where your process needs reinforcement - because the lifecycle acts as your guide.

That's why this book isn't about fixing one stage in isolation, it's about strengthening the full lifecycle so your planning process becomes a core part of how your organisation works, not something you constantly have to defend.

A structured lifecycle isn't bureaucracy, it's your operating rhythm, the foundation of reliable maintenance, and the signature of a planner who's running the system - not being run by it.

Key Takeaways

- Work orders are the core mechanism for managing maintenance activity in a CMMS.
- They exist to control, communicate, document, and measure work.
- There are multiple types of work orders, each with different urgency, rules, and structure.
- The full lifecycle includes request, triage, planning, execution, and close-out.

- Planners are best positioned to own the lifecycle and ensure it flows properly.
- Viewing work orders as a lifecycle builds better structure, discipline, and results.

Planner Exercise

Take a few minutes to reflect on your current work order lifecycle approach and culture:

1. How clearly is the work order lifecycle defined in your organisation today?
2. Which stage of the lifecycle do you see breaking down most often?
3. How much ownership do you currently take across each lifecycle stage - and where could you take more?

Use the next space to jot down your notes and thoughts before going onto the next chapter.

Chapter 2

Work Order Request Capture

Every work order begins with a request. Someone notices a problem, identifies a risk, or spots an opportunity to improve an asset, and flags it for action.

But here is an important truth: not every request should become a work order. In fact, many of them often shouldn't.

Some are duplicates. Some are vague or missing critical information. Others are just noise - personal gripes, speculative fixes, or non-maintenance issues being pushed through the CMMS out of habit or frustration.

As the planner, your job starts right here at the point of capture. If you let poor-quality input into the system, the downstream impact can be huge.

You'll spend more time chasing details that should have been included up front. Techs will arrive at jobs without the right information, tools, or access. Supervisors will lose trust in the plan. Eventually the whole WO process starts to feel like a revolving door of half-baked jobs and wasted effort.

A bad request pollutes the whole flow - and when this becomes the norm, it's planning that gets blamed. Not the person who raised the faulty request, but the planner who let it through. Because of that, this chapter focuses on the very first stage of the work order lifecycle: capturing usable requests and vetting them effectively.

You'll learn how to set clear expectations for requestors, how to spot red flags in low-quality submissions, and how to act as a strategic filter, not a pass-through administrator. Because what enters the system determines what comes out of it.

Great planners don't just manage the backlog, they protect the integrity of the system from the start.

What Makes a Work Request Usable?

Usable doesn't mean perfect. It doesn't mean the requestor needs to know every technical detail. What it does mean is the submission has enough clarity, context, and value to be triaged and planned.

A usable request should answer three key questions:

- **What's the problem?** (What do you see, hear, or suspect?)
- **Where is it happening?** (Asset, area, or specific equipment ID)
- **What's the impact?** (Is it affecting safety, performance, quality, or reliability?)

If those three questions can't be answered, the planner is stuck guessing - and that's how poor work orders enter the system.

What Do We Need - And Why?

At a minimum, a usable work request should include:

- Clear problem description
- Accurate location or asset reference
- Requestor contact info
- Any immediate access issues or production impacts
- Optional: supporting images, meter readings, or operator notes

Why does this matter so much? If a request is vague, the planner must chase down info. If the asset ID is missing, reporting links breakdown. If access or safety constraints aren't mentioned, the job gets rescheduled or dropped.

Good requests reduce rework. They let the planner plan. And they show the team that structure matters.

Setting Clear Expectations for Requestors

The planner's job isn't just to process requests - it's to shape how they're submitted.

That starts with:

- **Defined request forms** (digital or paper-based)
- **Mandatory fields** in the CMMS
- **Clear submission channels** for complete clarity
- **Controlled access** - not everyone should be able to raise a work order

Who should be allowed to submit requests? That varies. In some plants, only supervisors or leads can submit work. In others, operators are encouraged to flag issues directly. There's no one-size-fits-all - but whoever is involved must understand what "good" looks like.

Stakeholder Relationships and Support

You can't improve request quality in isolation. You need buy-in from other functions - especially operations.

Work request behaviour is often cultural. If operators think the planner ignores their tickets, they'll stop submitting. If supervisors think nothing gets prioritised, they'll bypass the process. That's why relationships matter.

Your role as a planner includes:

- **Coaching requestors** on how to submit clear, useful requests
- **Following up** on rejected or unclear tickets (and explaining why)
- **Training new staff** on the request process
- **Working with supervisors** to escalate or refine requests
- **Protecting the process** when others try to shortcut it

These aren't side jobs. They're core to controlling the front end of the lifecycle.

Planner Triage - Sorting the Noise

Once a request is in the system, it's your job to triage it. That means reviewing each submission and deciding whether it is:

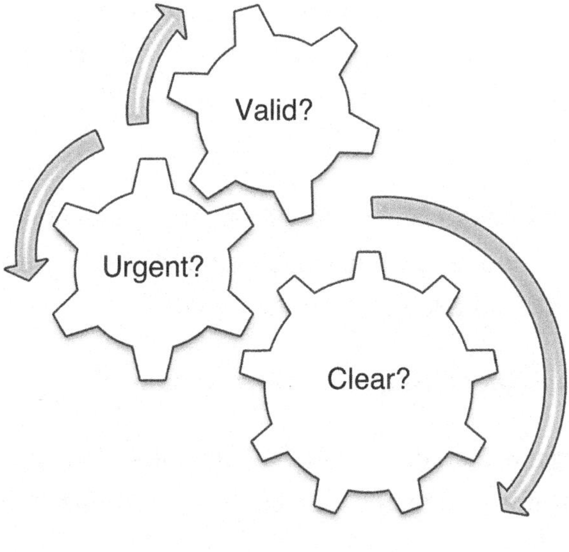

Triage is not just admin, it's your first and most important control gate.

When you do it well:

- Technicians don't waste time on vague or duplicate jobs
- The schedule reflects actual site priorities
- The CMMS data stays clean and consistent
- The team builds trust in the process

When you don't, work orders pile up, bad habits form, and the planner role becomes little more than a ticket processor.

The Importance of Acting as a Filter

Triage is how you protect the plan - and the people doing the work.

Every unfiltered request that makes it into the weekly schedule eats up resources. Every poorly scoped job creates confusion. Every vague or unnecessary task demotivates the team.

A good planner is firm but fair in this process. You're not there to say "no" to work - you're there to say "not like that, like this". Rejecting or returning a request isn't being difficult - it's being professional.

When done respectfully and with clarity, request rejections build credibility. People start to understand that not all work is equal, and that structure leads to smoother delivery.

When to Reject a Request - Strategically

Rejecting a work request can feel uncomfortable - especially when it comes from someone senior, vocal, or frustrated. But done properly, strategic rejection is one of the most important tools in your planner toolkit.

You're not rejecting work to block progress. You're rejecting it to improve the quality of what enters the system - and to protect the people and resources responsible for doing that work.

Poor requests lead to poor work orders. And poor work orders lead to poor planning, scheduling, and execution. So if you want better results downstream, you must take control upstream.

Reasons to Reject or Return a Request

Requests should be rejected, returned for rework, or placed on hold if any of the following apply:

- **No clear problem description -** "Machine not working" isn't enough. You need symptoms, context, or observed behaviour - not just the outcome.
- **No asset or location assigned -** If you can't link the job to an asset or work area, you can't plan or track it properly.
- **Duplicate request already exists -** Always check for existing work orders before accepting. Duplicates waste resources and undermine confidence in the system.
- **Unauthorised requestor -** If your site uses controlled access, ensure only approved personnel can submit certain types of requests.
- **No impact, urgency, or justification given -** If there's no reason to act, the job may not need doing - or not yet. Impact must be explained, not just implied.

The Right Way to Reject

When rejecting or returning a request, it's essential to:

- **Document the reason** in the CMMS, ticket, or tracker
- **Be specific and constructive** - point out what's missing or unclear
- **Offer a path forward** - e.g. "Please add location and re-submit"
- **Be consistent** - don't reject one person's request and accept the same issue from someone else

Where possible, follow up with a quick chat, email, or comment. This builds trust and encourages people to try again and work with you.

Building Trust Through Rejection

It sounds counterintuitive, but rejecting work requests - when done fairly and consistently - is one of the most powerful ways a planner can build trust across the organisation.

Why? Because it signals that there's a standard. That not just anything can enter the system. That requests are reviewed, filtered, and improved before becoming scheduled work. And that your time, and your team's time, is valuable.

When you set and hold that bar, people begin to understand what "good" looks like. They start submitting clearer requests. They stop seeing the CMMS as a dumping ground for every minor gripe. And they begin to see you - the planner - as someone who's not just pushing paper, but protecting time, people, and priorities.

Over time, those small rejections when paired with honest feedback, help shape a culture of better input and better planning. It becomes less about who shouts the loudest, and more about what work is genuinely worth doing.

That's how:

- Standards improve
- The system becomes more reliable
- The planner becomes a respected gatekeeper

Of course, rejection shouldn't be personal. And it should never come off as dismissive. The tone matters as much as the decision.

Because strategic triage isn't about saying no for the sake of it, it's about saying yes, once it's right.

And when people know that you'll only let through quality work - they trust that the system is worth using. That's how planning moves from being a box-ticking exercise to being a real driver of performance.

What to Look For - Common Gaps

Some common red flags during triage:

- "Not working" with no further detail
- No asset ID
- "ASAP" but no justification
- Overuse of emergency/breakdown priority
- Vague references like "inbound area" or "equipment in back"

With experience, you'll spot these patterns quickly. Over time, your goal is to reduce their frequency through coaching, feedback, and better front-end controls.

Practical Triage Workflows

Every organisation needs a defined work request triage process - even if it's simple. Here's a typical flow:

1. **New request enters system**
2. **Planner reviews daily**
3. **Categorise**: Emergency, corrective, planned, or invalid
4. **Validate**: Is info complete? Is the request justified?
5. **Decision**:
    - Accept and proceed to planning
    - Reject with reason and feedback
    - Hold pending more info

This basic flow works across industries - whether you're managing a 10-person workshop or a global CMMS with multiple planners.

What really matters is that the planner controls the gate.

Key Takeaways

- The quality of a work order starts with the quality of the request
- Usable requests are clear, specific, and submitted by authorised people
- Planners must actively triage, filter, and protect the flow of work
- Strategic rejection and coaching builds long-term credibility

- A simple, visible triage workflow helps keep the system clean and consistent

Planner Exercise

1. How would you rate the quality of work requests at your site today?
2. What's one change you could make to reduce low-quality requests or improve triage?
3. Who do you need stronger relationships with to make this part of the lifecycle more effective?

Use the next space to jot down your notes and thoughts before going onto the next chapter.

Chapter 3

Making Work Orders Schedule Ready

A well-run maintenance week doesn't begin on Monday morning - it begins the week before, when the planner ensures that every job on the schedule is ready to go – see the first Planner HQ Guide To: Mastering The Weekly Maintenance Planning Cycle for loads more detail on that point.

But it is the work readiness is the difference between a calm, structured week… and a chaotic, reactive one.

It's not the size of the backlog that breaks a plan. It's not always labour shortages or emergency breakdowns. More often than not, it's work orders hitting the schedule before they're ready. Missing parts, no clear job steps, permits not arranged, scoping left half-done.

Too many teams treat the schedule like a puzzle to solve at the last minute, but great planners know better. They don't just fill boxes, they build momentum upstream by making sure every job that gets scheduled is truly ready for execution.

That's where the leverage is. When you focus on the front end - on work order readiness - you create space for better planning, more accurate scheduling, and smoother delivery.

This chapter is about defining what "schedule-ready" really means, what it looks like inside your CMMS, and how you can build this discipline into your planning rhythm - so the work your team sees on Monday is clear, executable, and locked in with confidence.

Because great planners don't just push jobs into the system, they prepare their conditions for success.

A schedule-ready work order is:

- Fully scoped (you know what the job involves)
- Properly resourced (you've confirmed labour and parts)
- Safe to execute (permits or risk controls are defined)
- Clear in its instructions (technicians won't need to guess)
- Supported by operations (access, downtime, or handoffs are planned)

If any of those pieces are missing, you don't have a ready plan - you have a wishlist.

Key Checks: What to Look For Before Scheduling

Here's a practical breakdown of what every planner should be looking for when prepping a job for the schedule:

- **Parts confirmed**: Do you have everything needed on hand or on order?
- **Labour assigned**: Do you know who's available and capable to complete the job?
- **Access cleared**: Has downtime, isolation, or area control been discussed with operations?
- **Safety risks noted**: Are permits needed? Has the task been risk-assessed?
- **Instructions or job steps attached**: Even if brief, is there guidance to avoid improvisation?
- **Duration set**: Do you have a realistic time estimate, including setup and walk time?
- **Tooling or lifting needs**: Any special equipment required?

Not all of these checks will apply to every job - but every job needs some form of validation before it can be locked into the weekly schedule.

How This Prep Improves Planning Efficiency

Spending time on work order readiness might feel like a slowdown to the work. After all, there's always pressure to move jobs through the system, get the schedule filled, and show progress. However, this is one of those areas where slowing down actually speeds everything up.

When you take the time to properly confirm that a job is ready - scoped, resourced, safe, and clear - the entire week runs smoother. That investment of time upfront pays off in several key ways:

- **Fewer mid-week disruptions**: Jobs don't get cancelled last minute because of missing parts or access issues. Planners aren't scrambling to patch holes in the schedule with whatever is lying in the backlog.
- **Higher technician productivity**: Techs can start work without confusion. They don't waste time chasing down instructions, permits, or tools. They're able to focus on execution, not problem-solving.
- **Less fire-fighting for planners and supervisors**: When readiness is handled properly, there are fewer surprises. You're not spending your week untangling chaos - you're managing a process that's already been set up for success.
- **Better metrics and reporting**: When jobs are actually done on time and as planned, your completion rates, schedule adherence, and reliability data all improve. It

becomes easier to trust the numbers - and easier to use them to drive decisions.

- **Increased confidence from other departments**: Operations, leadership, procurement - they all notice when the maintenance schedule starts to become a tool they can rely on. It shows that planning is disciplined, structured, and adding real value.

This is where planning shifts from being reactive admin to being a driver of performance and execution. You're no longer just trying to stay on top of the chaos - you're building a system that prevents the chaos in the first place.

Over time, the impact is measurable. Smoother weeks, fewer carryovers, more engaged technicians, and a reputation for professionalism that sticks.

Common Red Flags That Delay Execution

Even small gaps can derail a scheduled job. Watch out for:

- Parts marked as "on order" but no ETA
- Work planned during peak production without confirmed access
- No one listed as responsible for isolations or permits
- Conflicting jobs scheduled on the same asset or in the same area
- Task instructions that assume tribal knowledge
- A "1-hour" job that actually takes 5 - because no prep was done

These are the blockers that cost credibility. They make you redo the schedule, cancel tasks, or send techs back to the shop empty-handed.

Your job is to spot them early and fix them before the schedule goes live.

Tying in Job Plans - Without Rewriting Them From Scratch

You don't need to reinvent the wheel every time.

If your CMMS supports job plans, use them - especially for routine or repeatable work. These templates save time, standardise instructions, and reduce errors.

Only apply standard job plans to jobs that are truly repeatable:

- PM routines
- Lubrication tasks
- Inspections
- Common repairs

For one-off or ad hoc requests, job plans are less useful. These require specific scoping - often with a planner or technician walking the job and making notes.

Use job plans as a base. Always review them to ensure they still apply. Tasks change, conditions change and assets change, so don't apply templates blindly.

Who to Involve in WO Readiness Prep

You don't need to do everything yourself - but you do need to own the process.

Depending on the job, you might loop in:

- **Reliability engineers** - for complex jobs or known chronic issues
- **Procurement or stores** - to confirm parts availability or kit build
- **Operations** - to lock in access, isolations, or support
- **Technicians or supervisors** - to confirm scope, safety, and sequencing
- **Safety reps or permit authorities** - if special controls or compliance rules apply

The key is not to try and solve everything solo. Your role is to coordinate, not control.

Ask the right questions. Surface the right blockers. Confirm the right inputs. Then you can schedule with confidence.

Building Your WO Readiness Checklist

Every planner should have a personal or team-level WO readiness checklist - a simple set of checks that any job must pass before it's allowed into the weekly schedule.

Here's an example to get you started:

- Scope is confirmed
- Labour assigned and available
- Parts available and/or staged
- Safety risks identified and permits arranged
- Instructions/Job Plan attached
- Estimated duration agreed
- Operations or access confirmed

You can adapt this list to suit your site - and over time, refine it based on what trips you up most often.

Making It Part of Your Weekly Rhythm

The best checklists in the world don't help if they're not used. Printouts stuck on the wall or tucked into a drawer won't change outcomes. To make WO readiness meaningful, it needs to live inside your weekly cycle.

That means carving out the right time - and protecting it - to do this properly. It's not just a task on the to-do list. It's a core part of planning.

Here's a practical flow to build that rhythm:

- **Review new work orders daily**: Spend 15–30 minutes each morning scanning what's come in. Categorise them quickly - ready to plan, needs more info, or not valid. This avoids build-up and keeps your workload visible.

- **Block out time mid-week for preparation**: Tuesdays and Wednesdays are ideal to start prepping jobs for the week after next. Confirm parts, check labour capacity, update durations, and chase any open questions. Don't leave it all until Friday.
- **Use Thursday as your final checkpoint**: This is your time to double-check the critical or complex jobs before freezing the schedule. Are the big-ticket tasks truly ready? Are any kits still in staging? Is operations aligned?
- **Freeze the schedule once readiness is confirmed**: Don't lock the plan because the clock says so - freeze it once you've confirmed that the jobs are deliverable. One missing permit or unavailable part can unravel the whole thing.

You're not just preparing work, you're preparing the conditions for successful execution. That's a big difference, and it's what separates admin-style scheduling from professional maintenance planning.

A planner who consistently delivers schedule-ready work earns trust quickly. The team knows that if it's on the plan, it's ready to go. And that kind of trust is gold.

Because in this role, your value isn't just about filling the calendar. It's about enabling work to happen properly, the first time. That's what makes a planner irreplaceable.

Key Takeaways

- "Schedule-ready" means the work can be done without delay, confusion, or rework
- Planners should confirm parts, labour, safety, scope, and access before scheduling
- Job plans can help, but must be applied wisely - not blindly
- A checklist or planning gate improves quality and consistency
- WO readiness should be part of your daily and weekly routine - not a last-minute scramble

Planner Exercise

1. What percentage of jobs on your current weekly schedule are truly "schedule-ready"?
2. What's the most common blocker or gap that causes jobs to be pushed or delayed?
3. What 3–5 checks would you include in your own WO readiness gate?

Use the next space to jot down your notes and thoughts before going onto the next chapter.

Chapter 4

Linking, Grouping and Sequencing Work

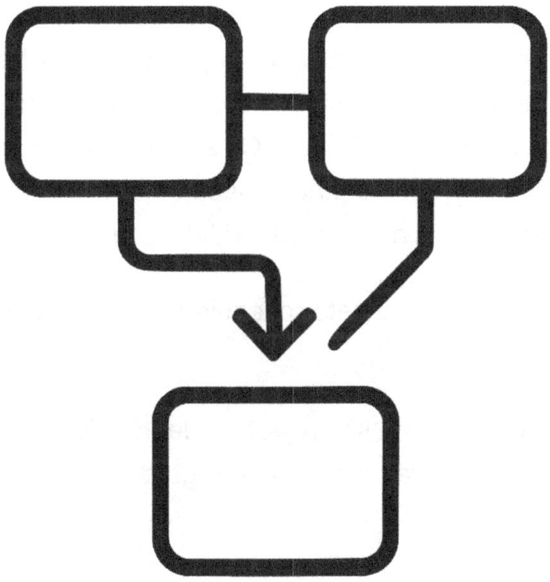

Work orders don't exist in isolation. One task often connects to another; whether it's a prerequisite inspection, a follow-on repair, or a multi-trade job broken into parts.

As a planner, it's easy to get caught managing each work order as a standalone item. In some cases, that's appropriate, but in many situations, true planning value comes from how you connect those pieces together. By linking, grouping, and sequencing related work, you create flow, you reduce waste, and you build schedules that are smarter, not just fuller.

It's not just about scoping a task or assigning labour. It's about recognising relationships between jobs and shaping the work to match the constraints and opportunities of your site.

That might mean grouping tasks by asset or location to avoid repeat downtime and repeat visits. It might mean sequencing a multi-trade job to ensure the electrical work happens before the mechanical fit-out. It might mean identifying follow-on repairs from PMs and linking them back for audit trail. Or splitting one giant work order into smaller, more manageable pieces that align with trade responsibilities and can fit around Ops with less disruption.

Great planning isn't just about what gets scheduled. It's about how that work is structured.

Let's explore the strategies behind smart work order relationships - and how to apply them in a way that works, regardless of your CMMS setup.

Understanding Work Order Relationships

Most maintenance tasks live within a network of related jobs. Understanding these relationships allows you to plan smarter and execute smoother.

Work orders can be related by:

- Asset: Multiple jobs on the same equipment
- Location: Work within the same general area or requiring the same access permit
- Trade: Mechanical, electrical, instrumentation, etc.
- Priority: High-risk or urgent work that needs bundling for visibility
- Condition: PM-triggered follow-ups or inspections that feed into repairs

Ignoring these relationships leads to inefficiency. You might send two techs to the same pump on the same day - without knowing the work could have been grouped and done together. Or perhaps worse, you schedule work without realising it depends on another task being completed first.

Parent and Child Work Orders

In most CMMS platforms, you can establish parent/child relationships between work orders. This means one “parent” WO represents the overarching task, and several “child” WOs break it down into specific activities - often by trade or subtask.

Example:

- Parent WO: Replace conveyor belt motor

- Child WO 1: Isolate and disconnect electrical feeds (Electrical)
- Child WO 2: Remove old motor and install new one (Mechanical)
- Child WO 3: Perform test run and restart sequence (Ops / Reliability)

This allows you to:

- Assign tasks to the right people
- Plan different durations or skill levels per task
- Track progress more accurately
- Avoid overwhelming one technician or team with a vague, oversized job

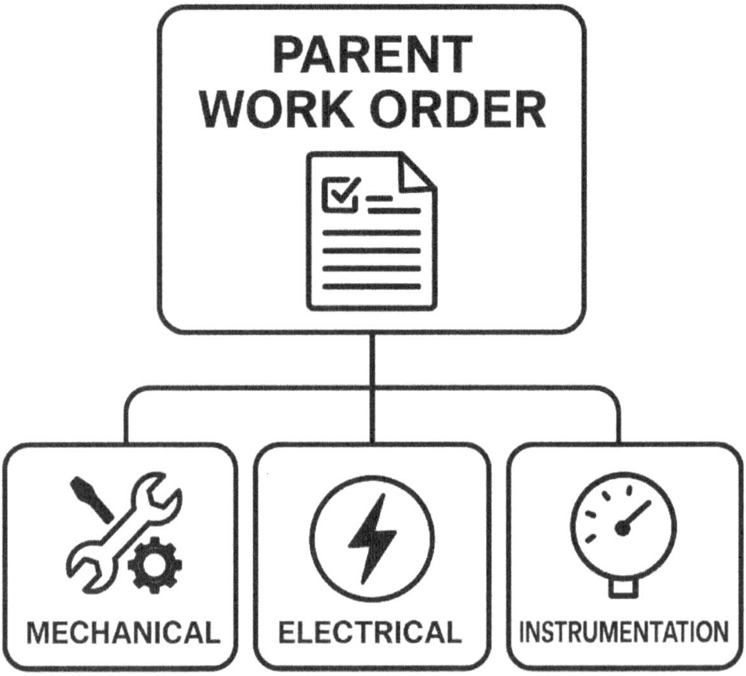

Splitting Work Orders for Clarity and Control

There's a common trap in planning: using one large work order to represent a complex job involving multiple trades, tools, and time slots.

While this might seem simpler from a tracking perspective, it often leads to confusion. Techs don't know which part they're responsible for, close-out is inconsistent, time tracking becomes messy and the job tends to drift without clear ownership.

Instead, split the work. Use your CMMS to create logically separated tasks (grouped under a parent, if needed) that clarify who does what, when, and how long it should take.

This gives you:

- Better control over labour
- More accurate close-out and time records
- Flexibility to reschedule parts of the job if needed
- Easier post-job review and improvement

When and How to Link Work Orders

Work order links aren't just about hierarchy. Sometimes they represent logical or operational relationships - like dependencies or sequential steps.

You should consider linking WOs when:

- One job must be done before another can begin
- A PM leads to a follow-on corrective task
- Work is split across time or trades but shares context
- You need to show a full chain of action for audit or analysis

Keep the method simple:

- Use your CMMS' built-in linking or relationship fields
- Add references in WO descriptions if needed (e.g., "See WO #12345")
- Maintain naming or tag conventions to flag linked work clearly

The goal is not to overcomplicate but rather to make related work traceable, understandable, and easier to manage.

Grouping Work for Efficiency

Grouping is about optimising the opportunity window. When you get access to an asset or area, you want to make that time count.

Group work orders when:

- Multiple jobs need the same access or isolation
- Tasks are on the same asset or system
- Work can be performed by the same technician or trade in one visit
- There's an opportunity to reduce setup, travel, or coordination effort

This is especially powerful when planning for:

- Outages or major inspections
- Contractor visits
- Shared labour windows
- Remote site trips

Before you finalise a weekly schedule, run a quick backlog review by area or asset. Look for jobs you can pull forward to maximise the access window.

Selling the Benefits to Operations

Operations often see maintenance as a disruption. But grouping and linking jobs effectively can help turn that perception around.

When you can say:

"We'll only need one isolation."
"We'll do everything on that line in one visit."
"This batch of work will reduce total downtime by 40%."

..You change the conversation.

The key is to frame the benefit in operational terms:

- Less downtime
- Fewer handovers
- More structured execution
- Predictability and reduced impact on production

Do this consistently, and operations will start bringing you in earlier. They'll see the value in planned maintenance. And you'll build the influence that every planner needs to make their system better over time.

Sequencing Work for Smarter Scheduling

Finally, sequencing. In more complex jobs with multiple steps - especially multi-trade or high-risk work - sequencing matters.

You need to define:

- What happens first
- Who hands off to who
- What must be verified before the next step starts

This can live inside a single work order (via job plan steps/activities), or be managed across linked WOs. The key is clarity.

Good sequencing avoids:

- Idle time
- Technicians waiting on each other
- Work being done out of order
- Rework or incomplete close-out

And downstream, it makes scheduling easier. You know what resources are needed, when. You can avoid bottlenecks. You can build time buffers where needed.

Key Takeaways

- Work orders often relate to one another - by asset, trade, or task
- Use parent/child structures or linking features to manage complex jobs
- Split large jobs into smaller, owned work orders to improve clarity
- Group tasks to optimise access and reduce repeat downtime

- Sequence work carefully to avoid delays and execution gaps
- Selling these techniques to operations strengthens long-term trust

Planner Exercise

1. Think of a recent job that involved multiple trades or steps. How was it structured? Could it have been clearer with linked or split WOs?
2. Look at your current backlog - are there grouping opportunities you've missed?
3. How would you explain the benefit of WO grouping or sequencing to your operations counterpart?

Use the next space to jot down your notes and thoughts before going onto the next chapter.

Chapter 5

Execution Week: Supporting Live Work Orders

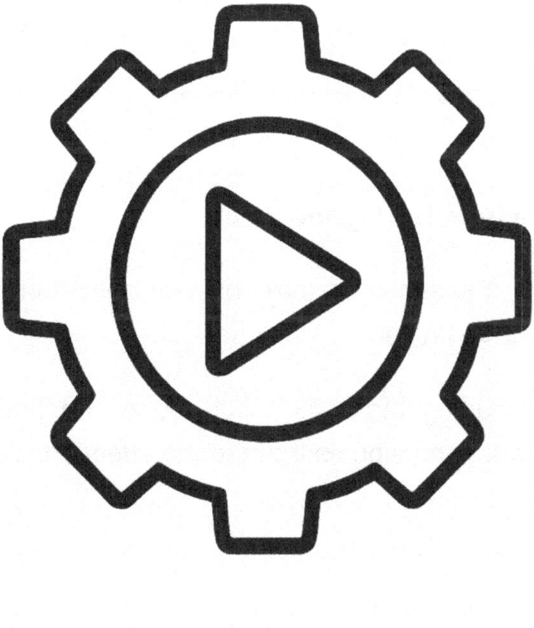

The schedule is live. The plan is published. And the work is underway. But for the maintenance planner, the job isn't done – it has shifted.

Execution week isn't about sitting back and watching things unfold. It's about monitoring, supporting, and adjusting the plan with discipline and intent. Because no matter how good your planning is, some things will always shift during execution. A critical part doesn't arrive, a technician calls in sick, a high-priority breakdown interrupts the plan.

This is where strong planning habits meet real-time responsiveness; and when done well, execution week becomes a time of stability, not stress. You don't need to control everything, but you do need to remain visible, engaged, and ready to help the team succeed.

This chapter is about how to do that effectively. What to watch, who to engage with, and how to protect the integrity of the plan - even when things change.

What the Planner Monitors During the Week

The planner's role during execution is part observer, part support system, and part quality controller.

You're not in the field doing the work - but you're constantly scanning for issues, gaps, and signals that require attention. You should be monitoring:

- Performance against the published schedule
- Daily work order progress (via CMMS or technician feedback)

- Unexpected delays, access issues, or resource shortages
- Completion status and time tracking accuracy
- Communication between maintenance and operations

This often involves a mix of digital oversight and boots-on-the-ground presence. As the planner, you don't need to shadow every job or hover over technicians, but you do need to stay connected to the feel of the week.

- **Checking in with supervisors and techs regularly**
  These check-ins don't have to be formal meetings. A quick conversation, a stand-up at the whiteboard, or a five-minute catch-up can give you a real sense of where things stand. Supervisors appreciate when you're engaged in the day-to-day flow, and technicians value knowing the planner cares about how jobs are actually progressing - not just how they look on the screen.
- **Walking the floor to stay connected to what's happening**
  A few minutes out of your chair pays huge dividends. Seeing jobs in progress lets you spot practical barriers: tools missing, access issues, or waiting on permits. It also builds relationships. When people see you in the work areas, they're more likely to share blockers or raise issues early, instead of letting them fester until the end of the week.
- **Reviewing job progress via WO status changes in the system**
  Your CMMS is your dashboard. Jobs should be moving through statuses: Scheduled → In Progress → Complete. Stagnant statuses are a red flag. A job stuck in "in

progress" for three days might be genuine, but it could also mean someone forgot to update, and both of those matter. Checking the system daily ensures the schedule reflects reality, not wishful thinking.

- **Noticing trends or blockers that repeat across different jobs**
  One delay is a hiccup. Five delays for the same reason is a pattern. Whether it's parts not arriving on time, permits holding things up, or operations access being denied, these recurring issues are opportunities for improvement. Keep a simple log or tracker so you can highlight these trends later - ideally feeding them into backlog reviews or continuous improvement discussions.

The goal isn't to micromanage, second-guess, or interfere with execution. It's to keep your finger on the pulse and provide support where you can add value. A planner who spots blockers early, communicates them clearly, and helps remove them quickly becomes a trusted partner - not just an administrator. And that's the difference between planning as a process and planning as a profession.

Managing WO Status, Delays, and Communication

A good planner tracks more than just what's on the schedule. You're not simply checking boxes, you're monitoring how the work is moving, where it's getting stuck, and what needs support to keep the week flowing. Execution is rarely perfect, but with the right tools and habits, you can keep momentum without sliding into firefighting.

Three tools make the difference during execution week: status updates, hold codes, and transparent communication.

1. Work Order Status Updates

Statuses are the pulse of the schedule. If you want to know how the week is going, don't just look at the schedule - look at the statuses.

- Ensure work orders are actively moving through the lifecycle: scheduled → in progress → complete. If jobs stall in one state for too long, it's a sign of a blocker.
- Look for patterns: Are many jobs sitting in "in progress" with no close-out? Are PMs staying in "scheduled" while breakdowns jump the queue? These signals help you spot stress points.
- Don't be afraid to nudge for clarity. A quick check-in with a supervisor or tech can reveal whether the delay is real or just a missed update in the CMMS. Both matter, because visibility depends on accurate status movement.

When statuses are updated consistently, you gain a near real-time view of progress. When they aren't, you're working blind.

2. Hold Codes or Delay Reasons

Not every delay is avoidable. But every delay should be understood.

Most CMMS platforms include hold codes - reasons you can apply when a job is paused. These add huge value when used properly:

- Common hold codes include waiting on parts, access denied, awaiting operations approval, safety permit pending, or resource unavailable.
- Hold codes give you a quick picture of what's really driving delays. If "waiting on parts" shows up 20 times in one week, you've just found a major opportunity for improvement.
- If your CMMS doesn't support this feature, track it manually. A simple planner log or delay tracker can capture reasons and feed them into your weekly review.

The point is not just to flag a delay, but to learn from it. If you know why work was held, you can work with stakeholders to prevent the same issue next time.

3. Transparent Communication

Information is only useful if it's shared. A planner's credibility during execution week comes from being clear, consistent, and proactive in communication.

- Keep a running note of delays or blockers. It doesn't have to be fancy - even a simple tracker can serve as a reference for daily updates.
- Share what's happening with the right people at the right time. If operations are waiting on a repair, they don't want

to hear about it three days later. They want to know now with context and an update on next steps.

- Be deliberate with your tone. Make it clear that you're tracking delays to improve the system, not to assign blame. When people trust that you're learning, not policing, they're more likely to keep you informed.

When you combine these three tools, you gain visibility, accountability, and a clear record of what happened during the week. This not only supports smoother execution in the moment, but also creates the feedback you need to plan better in the future.

This data feeds future planning. Patterns emerge. Root causes become visible. And improvement becomes possible.

Adding Reactive Work Without Breaking the Plan

Stuff happens. And sometimes you need to slot in new work mid-week. That doesn't make your schedule a failure - as long as it's done with discipline.

Before adding anything:

- Ask: Does it need to be done this week?
- Check: Is there room without bumping other work?
- Decide: Can it be absorbed into a current WO, or does it need a new one?

When done right, adding reactive work can show maturity in the planning process. You're not rigid, you are structured and flexible.

If you add everything that gets flagged, your schedule becomes meaningless. It's a to-do list, not a plan.

A few principles to work by:

- Protect the "frozen" core of the schedule. Only adjust outer layers.
- Keep a visual or digital board showing new work vs. planned work
- Communicate what has changed - and why
- Flag any carryover risks that result from mid-week additions

Doing this well builds trust with operations and techs alike. They know you're responsive but also have clear boundaries and structure.

Maintaining Visibility and Discipline Mid-Week

Some planners retreat during execution. Once the schedule is out, they disappear into next week's planning, CMMS admin, or other tasks.

This is very understandable, but is also when problems often creep in. Your mid-week visibility matters. Not just for monitoring, but for morale. The team sees you're engaged and that the current plan still matters.

You don't need to be on the tools or in every meeting. But:

- Walk the floor once or twice a week
- Ask techs how jobs are going

- Drop into toolbox talks occasionally
- Join huddles or handovers if useful
- Respond quickly to questions or requests

This builds planner–technician relationships that pay off in future weeks. You get better feedback. More heads-up about issues. And more willingness to follow the plan.

Avoiding the "Black Hole" Effect

One of the most common failure modes in maintenance planning is the black hole effect.

The plan is published. And then... nothing.

No updates. No follow-up. No idea how the work is going.

Work orders sit untouched. Completed jobs don't get closed out. Issues go unreported. And no one's sure who's watching.

As the planner, you must keep the loop alive.

- Track what's happening
- Ask for updates
- Log issues and blockers
- Prep for your Friday review by midweek
- Keep visibility high

Otherwise, all the work you did to create a good schedule becomes lost in execution.

When people see you staying active through the week, the system gets stronger. The schedule stays relevant. And the planner role becomes a real asset - not just a document generator.

Key Takeaways

- The planner's job doesn't stop once the schedule is published
- Monitor job progress, status changes, and delays with intent
- Use hold codes or delay logs to track common blockers
- Add reactive work strategically, without undermining the plan
- Maintain visibility mid-week - the team notices
- Avoid the "black hole" effect by staying engaged from Monday to Friday

Planner Exercise

1. How do you currently track job progress during the week? Where are the gaps?
2. What could you do mid-week to increase your visibility and impact?
3. Think of a recent week where a reactive job was added - how did it affect the plan?

Use the next space to jot down your notes and thoughts before going onto the next chapter.

Chapter 6

Closing Out Work Orders with Discipline

There's a moment of relief when a job is done - the equipment's running again, the team moves on, and the pressure fades. But for the planner, the work order isn't truly finished until it's been closed out correctly.

Closing out isn't just an admin step, and that is a common misconception. Closing out is where the work becomes usable data. It's the handoff between execution and learning and is what turns "we did the job" into "we know what happened, why it happened, and how it went."

Yet in many organisations, this part is neglected. Technicians race to the next task. Supervisors forget to check the system. Planners are left chasing missing fields and guessing at job outcomes. And slowly, the CMMS fills up with noise; incomplete records, unreliable metrics, and missed opportunities to improve.

A good close-out process creates the foundation for all future planning, reporting, and reliability improvement. A weak one - silently undermines everything you're trying to build.

This chapter is about how to build discipline around work order close-out, not as an administrative burden but as a vital part of the planning cycle.

What Good Close-Out Looks Like

A closed work order should tell a complete, accurate story of what happened in a way that supports future analysis, compliance, and planning.

Depending on the type of job, that might include:

- Actual start and finish time: Helps validate job duration and scheduling accuracy
- Booked labour hours: Who worked on it, for how long
- Spare parts consumed: Pulls from stock and informs reordering
- Failure codes: For breakdowns or defects, indicating failure mode and root cause for future trending
- Technician notes: What was done, how it was done, and any issues found
- Follow-on work identified: If more work is required, it gets captured
- Completed checklists or inspections: For regulated or PM work
- Close date and responsible person: So the record has an accountable ending

For proactive work (like PMs or inspections), the close-out should also flag:

- Was the equipment found within tolerance or spec?
- Were there early signs of wear or risk?
- Was the task completed fully, or deferred for follow-up?

For reactive work, the detail is even more important. This is your post-mortem - and your chance to stop future failures.

The Planner's Role in Coaching Good Habits

You're not the one closing every work order - but you are usually the first person to notice when it hasn't been done properly. A missing failure code, vague notes, no actual hours booked - these gaps all become obvious when you try to use the data for scheduling, reporting, or planning the next job.

And that puts you in a unique position. It gives you influence.

Your role isn't to quietly fix the data yourself or to grumble about poor habits. Your role is to coach others on what good looks like and why it matters. The quality of close-out affects the entire maintenance process. If technicians understand that, they're far more likely to treat it as part of the job rather than a box-ticking exercise.

Coaching in this context means:

- Giving feedback to technicians when notes are vague or missing. Don't wait until it's a pattern. A quick comment like, "Could you expand this so we know exactly what was replaced?" teaches more than a policy document.

- Explaining how their notes support planning, not just admin. Show them that good close-out makes it easier to order the right spares next time, set realistic durations, and identify recurring failures. Notes aren't just for the CMMS they're for the future of the job.

- Reminding supervisors of the bigger picture. Supervisors are the ones closest to the team. When you share how poor close-out damages reporting, audits, and improvement efforts, you equip them to reinforce expectations at the frontline.

- Leading by example. If your planning inputs are sloppy, no one will take your calls for discipline seriously.

Demonstrating accuracy in planning (durations, parts lists, safety notes) sets a clear standard that others will mirror in their close-out.

- Working with your CMMS admin or reliability team to simplify the process. If close-out screens are cluttered or confusing, no amount of coaching will fix it. Streamlining unnecessary fields or reordering entry steps can reduce frustration and improve compliance.

Coaching doesn't mean nagging or micromanaging. It means showing people that their small actions have large consequences. It's about helping them connect today's close-out to tomorrow's performance, and building pride in getting it right.

A planner who takes this role seriously shifts the culture. Over time, close-out stops being a chore and starts being part of the team's identity and basic routines.

The Supervisor's Role in Reinforcing Discipline

While the planner sets the expectation, it's the maintenance supervisor / team leader who should usually drive compliance.

They are the daily people managers. The ones closest to the techs, and they set the tone.

If the supervisor accepts sloppy close-out, the team follows suit. If they check, correct, and reinforce standards, it becomes culture.

Planners should work alongside supervisors, not in conflict:

- Share examples of good and bad WO close-out
- Highlight trends or gaps in the data
- Encourage positive reinforcement when good habits show
- Align on what "minimum data standards" look like for your site

Together, you create a system where good close-out becomes normal, not optional.

How Poor Close-Out Kills Analysis

When close-out data is poor, so is your visibility. And when visibility disappears, so does control.

Consider that the CMMS is only as good as the information it holds. If jobs are marked complete without proper detail, the system becomes an illusion of progress - work is "done" on paper, but the reality is far murkier. Decisions based on that data are, at best, unreliable and, at worst, completely wrong.

Here's what happens when work order close-out discipline slips:

- Missing failure codes mean recurring issues vanish into the noise. You can't spot patterns, and ongoing/repeat equipment problems stay hidden.
- No actual hours makes labour planning a guessing game. Schedules drift, utilisation is skewed, and you can't defend your resource capacity planning to leadership.

- No parts booking breaks your stock forecasting. Stores never knows what's truly being consumed, leading to overstocking some items and running out of others.
- No technician notes mean planners repeat the same mistakes. Every breakdown feels like the first time because no history is captured in usable detail.
- Work marked complete when it's not destroys trust in the CMMS. When operations sees a "completed" job on screen but knows the asset is still faulty, confidence in the entire system collapses.

These failures don't stay isolated. They build over time, feeding into a dangerous cycle of decline:

- Teams stop trusting reports and stop using them
- The CMMS becomes a dumping ground, not a useful management tool
- Planning reverts to reactive, because the system no longer supports proactive work
- No one feels ownership of the data - it's "someone else's job"
- Improvement efforts stall, because the foundation is rotten

The cost of poor close-out is hidden at first. But over months and years, it erodes the very reason you have a planning function. Without accurate close-out, the organisation loses the ability to learn, improve, and prevent repeat failures.

A strong close-out habit breaks that cycle. It creates a reliable record of work done, a trustworthy source of performance data, and a foundation for continuous improvement. It turns maintenance into a measurable, improvable system - not just a string of repairs and forgotten fixes.

Auditing Close-Out for Continuous Improvement

You don't need to audit every work order, but regular sampling and spot checks helps you notice patterns, coach better, and improve system performance.

Try:

- Spot-checking 5 closed WOs per week
- Sampling a mix of reactive and proactive tasks
- Looking for trends in missing fields or vague notes
- Sharing findings with supervisors and techs (positively)
- Feeding insights back into planning routines or job plan updates

Also look at big picture trends:

- Which techs close out consistently well?
- Are certain work types regularly missing data?
- Does close-out drop off on Fridays or busy weeks?

This isn't about blame - it's about building a better system together.

Key Takeaways

- A properly closed work order tells a complete story of what happened
- Essential fields include time, labour, parts, findings, and failure codes
- Planners help coach good habits and flag data quality issues
- Supervisors play a key role in setting standards and reinforcing discipline
- Poor close-out leads to data decay, unreliable reports, and repeat problems
- Regular audits help identify gaps and improve over time

Planner Exercise

1. How could you better support supervisors in enforcing data discipline?
2. Review 5 recently close Wos. What is missing? What has been well done?
3. What 3 things would you include in a "Minimum Close-Out Standard" for your site?

Use the next space to jot down your notes and thoughts before going onto the next chapter.

Chapter 7

Building a Sustainable Work Order Process

A well-run work order system isn't built in a day. It's built over time - through structure, ownership, and consistency. It's easy to focus only on this week's schedule, this month's backlog, or the latest urgent repair. But the real value of strong planning comes when you step back and think bigger.

- How do we make the work order process reliable - not just reactive?
- How do we ensure the system holds up when people change roles, when workloads spike, or when priorities shift?
- And how do we stop the system becoming bloated, bureaucratic, or ignored?

This chapter is about building a work order process that lasts. Not just for the planner, but for everyone involved. It's about embedding clear responsibilities, maintaining visibility, and continuously improving the process - without letting it become admin for admin's sake.

A sustainable WO process doesn't just support the work. It improves it.

Establishing Ownership Across Departments

One of the biggest failure points in any workflow is blurred ownership. When it's unclear who does what, things fall through the cracks.

The work order lifecycle involves multiple people:

- Requestors
- Planners

- Supervisors
- Technicians
- Operations
- Reliability or engineering teams

Unless roles and responsibilities are defined, each group assumes someone else will handle it.

Use a RACI model to map this out:

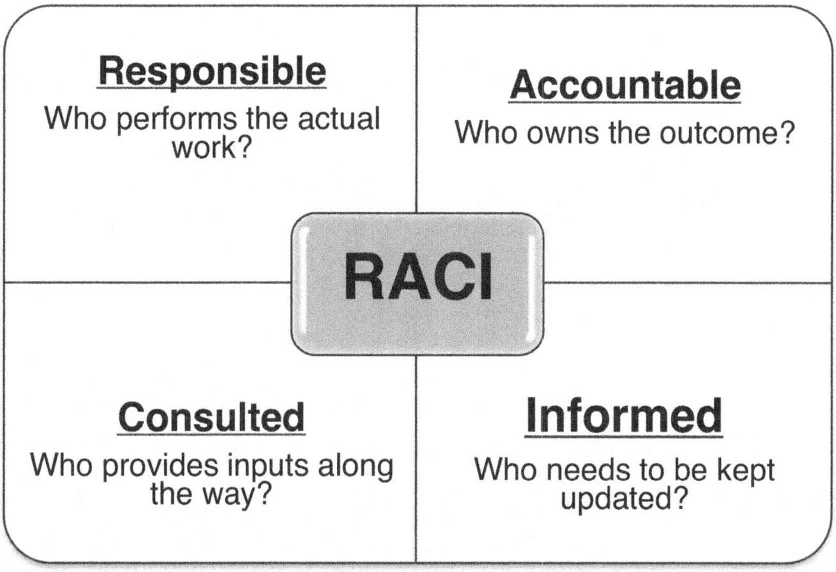

Example: For work order close-out:

- Responsible: Technician
- Accountable: Supervisor
- Consulted: Planner
- Informed: Reliability or Ops lead

Building this level of clarity removes ambiguity - and creates shared standards.

Standardising Lifecycle Expectations

Every site has its quirks. But if every planner or supervisor uses their own version of the work order lifecycle, things fall apart fast.

Sustainability requires shared understanding.

That means:

- Standard definitions of what each WO status means
- Clear expectations for what "ready" and "closed" actually include
- Shared checklists or gates for planning and scheduling
- Agreed rules for reactive work, follow-ons, and hold statuses

Document these expectations in a short, usable format - not a 40-page SOP no one reads. Use diagrams, quick guides, and examples to get people aligned.

And once you've defined the standard, train to it. New hires, contractors, and even cross-functional teams benefit when everyone understands how work flows.

Planner Health Checks: Auditing the System

As a planner, you're in the best position to spot when the process is slipping.

Build in mini audits - small, regular checks to test the health of the work order process. These might include:

- Weekly spot checks of closed work orders
- Review of “in progress” jobs with no updates
- Checking for WOs stuck in planning or draft stages
- Reviewing triage outcomes or rejected requests
- Spotting repeat issues in job scope, notes, or labour booking

Keep a simple log or tracker. Note what’s working, where gaps appear, and who needs coaching or support.

Over time, this becomes your feedback loop for improving the system - not just the schedule.

Using Dashboards, Reviews, and Trackers

You don’t need a fancy software system to track process performance. But you do need visibility.

Use whatever tools you have and whatever you can work with: CMMS dashboards, Excel trackers, whiteboards - to identify and visualise what is happening in the process.

You might track:

- Weekly WO throughput (opened, scheduled, completed)
- Overdue or stalled work orders
- Quality issues in WO close-out
- Number of requests rejected vs. approved
- Carryover trends by team or area

Visualising this builds engagement. It creates useful discussion points in planning meetings. And it turns planning into something that can be reviewed, discussed, and improved.

Keeping the Process Lean

One of the biggest threats to sustainability is complexity.

If your work order process becomes overloaded with admin, steps, approvals, or checklists, it risks people not using it. A good planner is able to simplify systems and processes and therefore improve engagement.

Watch out for:

- Overly long WO templates with fields no one uses
- Approval processes that delay urgent work
- Job plans that are copied blindly and never updated
- Excessive statuses or codes that confuse more than clarify

Your goal isn't to control every detail. It's to build just enough structure to create flow and accountability - without clogging up the work.

Every process has friction. Your job is to make sure it's the right kind of friction - the kind that prevents failure, not the kind that slows everything down.

Continuous Improvement That Involves Others

You see the patterns. You feel the pain points. But the process doesn't belong to planners alone.

The best improvements come when you involve others - techs, supervisors, ops leads, and even reliability or stores.

Try:

- Monthly process review huddles
- Data-sharing meetings with operations
- Short surveys to gather planner–tech feedback
- Co-creating checklists or gates with the team
- Publishing simple before/after examples of process fixes

This creates buy-in. People start to see planning not as red tape, but as a shared system to make everyone's work smoother and more effective.

Sustainability doesn't mean the process never changes. It means it improves without collapsing under its own weight.

Key Takeaways

- A sustainable WO process needs clear ownership across departments
- Use RACI models to define who does what at each lifecycle stage
- Standardise lifecycle expectations and train others to follow them

- Run regular planner health checks to spot data or workflow issues
- Visual dashboards and trackers create transparency and engagement
- Keep the process lean - add value without overcomplicating
- Involve others in continuous improvement to build long-term buy-in

Planner Exercise

1. What part of your current WO process feels too complicated or unclear?
2. Where is ownership vague - and how could a RACI help clarify it?
3. What's one small process fix you could test with your team this month?

Use the next space to jot down your notes and thoughts before moving on.

Conclusion

Wrapping Up The Lifecycle

If there is one thing to take away from this handbook, it is that you, the planner, own the flow and the work order lifecycle.

Not in the sense that you control every single step, but in the sense that you're the one holding the thread. From the first work request to the final close-out note, the work order flows through a series of decisions, handoffs, and updates. And while many hands touch the process, it's the planner who keeps the system connected.

That's the difference between reacting to maintenance work and managing it.

Owning the flow means making sure requests are usable. That the work is planned and ready. That the schedule is achievable, and the close-out is complete. It means having a watchful eye over the lifecycle - not just to move work along, but to improve how the work happens in the first place.

This is why your role is so critical. You're not just entering data or publishing schedules. You're enabling reliability. You're creating structure. You're setting a standard that ripples across departments.

Planners don't just work in the system. They shape how the system works.

From Firefighting to Framework

For too many teams, the work order process feels like chaos. Requests flood in. The backlog grows. Priorities shift. People act based on what's urgent, not what's right. And planners are left trying to make sense of the noise - chasing updates, rescheduling work, and explaining gaps that shouldn't have existed in the first place.

But there is another way, and it starts with viewing the work order as more than just an administrative form. It's a framework.

When you treat the work order lifecycle as a structured process - not a free-for-all you start to regain control. You stop scheduling guesswork, you improve technician handovers, and you build better data and stronger trust across the organisation.

A good framework doesn't remove flexibility. It gives you the confidence to adapt without losing clarity. It gives teams a rhythm to work within. And it gives the organisation a way to scale maintenance without scaling chaos.

Work orders are the foundation of the CMMS. But it's how you manage the process around them that determines whether your system works for you or just adds more admin to the week.

This book has given you a blueprint. Seven chapters as a complete system you can start refining and applying today.

Small Fixes, Big Impact

The good news is that you don't have to fix everything at once.

You don't need a full process overhaul or a new CMMS to see results. Most of the biggest wins come from small changes done consistently. And that's where planners shine - noticing the gaps, fixing the friction, and nudging the process forward.

Maybe it's improving the way you triage work requests, or maybe it's getting stricter about schedule readiness.

Maybe it's coaching one tech on better close-out notes and watching that improve reliability reporting three months later.

These aren't headline-grabbing changes. But over time, they add up. They build trust. They improve efficiency. They make you the person that other teams rely on - not just for answers, but for clarity.

That's the trait of a great planner: not perfection, but persistent improvement.

The lifecycle isn't just a process map. It's a daily opportunity to improve how your organisation works.

Your Next Steps

By now, you've seen the full picture of how a work order should flow - and where most systems break down.

If you're ready to make a difference, start small:

- Pick one part of the lifecycle that needs tightening
- Create a checklist or standard for that one step
- Get feedback from your team and refine it
- Monitor the results for 30 days and adjust again

Change doesn't have to be sweeping. It just has to start.

This handbook is part of the growing **Planner HQ ecosystem** - a practical, professional resource hub for modern maintenance planners. Alongside this series, you'll also find:

- The essential flagship book for all planners, *The Maintenance Planner's Playbook* - a deeper dive into the full scope of the planner role and how to succeed
- A growing book library covering specialist topics like job planning, spare parts, and CMMS best practices
- Free and premium content, downloads, and training tools designed to help you lead with more structure, strategy, and impact

Planner HQ exists for planners like you - the ones who want to move beyond the chaos and become a force of reliability inside their organisation.

Scan the QR code to head straight to the Planner HQ website

BONUS CHAPTER

5 Common WO Lifecycle Mistakes

(And How to Fix Them)

The work order lifecycle has plenty of opportunities for things to go wrong. Even experienced planners can fall into traps when the pressure is on, or when bad habits creep in. These mistakes aren't just small slip-ups - they can undermine the entire process, frustrate technicians, and erode confidence in planning. Let's look at five of the most common mistakes, why they happen, and how you can avoid them.

Mistake 1: Accepting Vague or Low-Quality Requests

Why It Happens

Requests come in from all angles. Operators, supervisors, and sometimes even technicians drop tickets into the CMMS, often with limited detail. Planners are under pressure to keep work moving, so there's a tendency to accept whatever lands and deal with the gaps later. It feels easier in the moment to "just let it through" rather than push back and risk conflict.

The Problem

When vague requests are accepted, the planner ends up doing detective work after the fact. Missing asset IDs, unclear problem descriptions, or no indication of urgency all slow down the process. The backlog grows with jobs that can't be prioritised properly. Worse, technicians arrive at jobs with no real guidance, leading to wasted time or unsafe improvisation. Over time, poor-quality requests clog the system and make planning look ineffective.

The Fix

Set clear standards for what constitutes a usable request. Minimum information should include asset or location, a clear description of the problem or need, and any immediate impact. If this isn't present, return the request with a note on what's missing. Support this by working with operations and supervisors to educate requestors on what "good" looks like. Over time, the quality of requests improves - not by chance, but because you've consistently held the line.

Planner Tip

Don't see rejection as blocking work. Frame it as protecting the team's time. The more consistent you are, the more people learn. Every vague request you let through adds to your chaos. Every vague request you reject - with feedback - builds a stronger system.

Mistake 2: Scheduling Work Before It's Ready

Why It Happens

There's pressure to fill the schedule. Managers want to see full utilisation. Operations push to have their requests included. Planners, eager to show progress, sometimes schedule jobs that aren't truly ready - parts not yet available, scope incomplete, or access not confirmed. It looks good on paper but sets the week up for failure.

The Problem

Jobs that aren't fully planned collapse mid-week. Technicians arrive on site and discover they're missing parts. Access hasn't been arranged, so the work is delayed. Supervisors shuffle jobs around, undermining the frozen schedule. The knock-on effect? Lower compliance, frustrated technicians, and reduced trust from stakeholders. Instead of demonstrating control, planning looks like guesswork.

The Fix

Introduce a readiness gate. Before any job makes it onto the weekly plan, confirm the essentials: scope defined, parts staged or ordered, permits in place, access confirmed, and labour available. If any of these checks fail, hold the job back until it's ready. It's better to run a smaller schedule that delivers than a full one that falls apart. Over time, stakeholders will value your consistency more than short-term promises.

Planner Tip

Don't let pressure dictate readiness. The frozen schedule is your product - your reputation. Protect it by ensuring every job you release to the plan has passed your readiness checks. If you wouldn't stake your credibility on it, don't schedule it.

Mistake 3: Treating Work Orders as Isolated Tasks

Why It Happens

With dozens or even hundreds of jobs in the backlog, it's natural to think in terms of individual tickets. The CMMS presents them as discrete records, so planners fall into the habit of managing them one by one. It feels efficient to just process the queue as it comes, without stepping back to see the bigger picture.

The Problem

When jobs are treated in isolation, opportunities for efficiency are missed. Multiple tasks on the same asset get scheduled separately, forcing repeat isolations or downtime. Multi-trade work ends up under one vague WO, creating confusion about who does what. Inspections that trigger corrective actions aren't linked, making it harder to track follow-ups. The result is wasted time, duplicated effort, and frustrated operations.

The Fix

Recognise relationships between jobs. Use parent/child structures to split complex tasks across trades. Group jobs by asset or area to reduce repeat setups. Link follow-on corrective actions to their original PMs. Build sequencing into multi-trade jobs to ensure logical flow. This doesn't just help the technicians - it improves schedule efficiency and strengthens the traceability of your CMMS data.

Planner Tip

Think beyond the ticket. Every time you touch a work order, ask yourself: does this connect to anything else? The more you plan relationships, the more efficient your schedule becomes.

Mistake 4: Weak or Inconsistent Close-Out

Why It Happens

Close-out often falls to the end of the day or the bottom of the priority list. Technicians are busy, supervisors are focused on tomorrow's work, and planners are under pressure to close jobs quickly for reporting. As a result, actuals, notes, and failure data are skipped or rushed.

The Problem

When close-out quality is weak, the system loses its memory. Actual durations are inaccurate, parts usage is unreliable, and failure codes are inconsistent. Without this data, planners can't estimate properly, reliability teams can't analyse performance, and leadership can't trust the reports. Over time, the CMMS becomes little more than a record of "done" rather than a source of improvement.

The Fix

Make close-out a non-negotiable standard. Set expectations with supervisors that work isn't finished until it's documented. Audit closed work orders weekly to check quality and provide feedback. Recognise and share examples of good technician notes. Show how close-out data improves job planning accuracy and strengthens PMs. When people see the feedback loop in action, they're more likely to take it seriously.

Planner Tip

Don't close jobs on behalf of others just to clear the queue. Hold the line. The integrity of your data is worth more than hitting an arbitrary number of "completed" jobs.

Mistake 5: Not Bringing Operations Into the Process

Why It Happens

Operations are often seen as the customer of maintenance - the ones who submit requests and receive work. Planners sometimes treat them as external rather than integrated, handing over a schedule with little discussion. With pressure to keep maintenance running, it feels simpler to keep the process within the maintenance team.

The Problem

When operations aren't engaged, they see maintenance as disruptive. Jobs are scheduled at bad times. Access isn't coordinated. Downtime clashes with production needs. The result is friction, resistance, and last-minute changes that undermine planning. Without trust and collaboration, the planner's role becomes harder and less respected.

The Fix

Bring operations into the conversation. Discuss downtime needs, group work to minimise interruptions, and show the benefits of efficient scheduling. Frame your planning in terms of operational impact - reduced downtime, fewer isolations, smoother execution. The more operations see the value, the more they'll support your process and provide the input you need.

Planner Tip

Don't wait for operations to come to you. Take the lead in building the relationship. Every time you demonstrate that planning helps production, you strengthen your influence - and your ability to run the lifecycle properly.

Planner FAQs:

Real-World Answers to Weekly Planning Challenges

?

Even with a solid process, questions always come up.

This section tackles the real-world challenges planners face once they start adapting their work order lifecycles.

Think of it as your quick-access troubleshooting guide for the messy, day-to-day reality of planning.

1. How detailed should a work order really be?

Enough to get the job done right - but not so bloated that it slows the team down. Focus on clarity of scope, clear steps if required, and any safety or asset-specific info that changes the way a technician would approach the work. Repeating the same background details on every WO is wasted effort. Prioritise relevance.

2. What's the best way to get better close-out notes from techs?

Two things: consistency and context. First, make it a habit. Review notes regularly and give feedback (both positive and corrective). Second, explain why the data matters. If techs understand that their notes drive root cause analysis, recurring work prevention, or performance tracking, they're more likely to take it seriously.

3. Is it okay to schedule work that isn't fully planned yet?

Only in controlled exceptions. If a task must be done but is missing a job plan, you might fast-track it - but make sure to communicate risks and flag it as needing follow-up. Don't let this become your norm. Your schedule is only as strong as the quality of what enters it.

4. How do I avoid chasing updates on dozens of in-progress jobs?

Build visibility into your routine. Use dashboards, daily checks, or reports to flag what's stuck. Group in-progress jobs by technician or area for quicker review. And engage with supervisors - they should own daily follow-up. Your role is to keep the process flowing, not to micromanage.

5. What should I do when work is completed but the WO hasn't been closed out?

Establish a time limit. Work that's physically complete should be closed out - or at least updated - within 24–48 hours. If not, follow up with the responsible person and look for patterns. Are close-outs delayed due to system access, unclear expectations, or low priority? Fix the root, not just the symptom.

6. Can I link multiple trades to one work order, or should I split it up?

Split it. You'll get better tracking, clearer accountability, and more manageable scheduling. Use parent/child relationships to keep everything tied together. Trying to manage electrical, mechanical, and contractor work all on one WO usually leads to delays, confusion, and reporting issues.

7. What's the planner's role when the team is short-staffed?

It's tempting to jump into coordination or even physical work - but your core responsibility is to protect the process. Prioritise work based on capacity, communicate delays early, and ensure that the system doesn't break down entirely. Helping out is fine - but not at the cost of losing planning control.

8. Do I need to audit completed WOs every week?

Not every single one - but regular spot checks are essential. Look for trends in poor quality (e.g. blank failure codes, inaccurate durations, missing notes). Use findings to coach individuals or improve processes. A few well-targeted audits each week can dramatically improve long-term data quality.

9. How do I deal with repeat work requests for the same issue?

Create a workflow for identifying duplicates - ideally in your triage process. Look for history on the asset and tag repeat requests to an open WO when relevant. If repeat work is valid, start a root cause conversation: Is the repair being done properly? Is the asset overdue for upgrade? Don't just keep closing the same job forever.

10. What if no one seems to care about the WO process but me?

You're not alone - and it's frustrating. Start by finding one ally: a supervisor, lead tech, or ops partner who understands the value of structure. Build small wins with them, then show others the results. Planning is often thankless until it fails - but over time, consistent structure speaks for itself.

11. What if a work order goes "missing" in the system?

It happens more than people admit. The job gets done, but the WO was never updated - or it never moved past "planning." The best defence is regular lifecycle status reviews and dashboards that flag stale WOs. If something hasn't moved in 7+ days, it needs checking.

12. How can I standardise failure codes and close-out fields without overwhelming the team?

Keep it simple. Use drop-downs, limited options, and clear definitions. Ask techs to focus on the top 2–3 fields that drive the most value - e.g. failure mode, actual time, and technician notes. Standardising everything at once causes friction. Prioritise what's most useful for your reporting and reliability efforts.

13. Who should "own" the CMMS - the planner or the system admin?

The system admin owns the infrastructure and permissions. The planner owns the workflow. You're the one making sure work flows properly from request to close. That means planners should lead on status flows, templates, job plan quality, and WO discipline - even if you don't own the backend config.

14. What should I do when someone bypasses the process and just "gets the work done"?

Treat it as a coaching opportunity, not an attack. Ask: Why did they bypass the process? Was it urgent, or do they not trust the system? Reassure them that the WO process exists to protect quality, traceability, and workload visibility. Show them how their actions ripple through the system.

15. Is it okay to bulk close work orders at the end of the week?

Only if you want unreliable data. Bulk-closing WOs often leads to inaccurate time booking, missed notes, and no insights into what really happened. Encourage real-time or daily close-out, and make it clear that speed isn't the goal - discipline is.

16. How should I handle work requests that aren't really maintenance?

Redirect them respectfully. Have a list of common non-maintenance categories (e.g. cleaning, IT, admin) and direct users to the correct route. If your CMMS allows, use a rejection reason like "out of scope" or "forwarded to appropriate team." Use these moments to protect the integrity of the system.

17. Can the planner be the one who creates the job plan?

Yes - but it depends on the site. If your site doesn't have a dedicated reliability engineer or planning standards team, the planner often takes the lead. Just make sure you have enough input from experienced technicians or SMEs to get the details right. Job planning isn't guesswork - it's technical storytelling.

18. What's the best way to teach others about the lifecycle process?

Use a simple diagram and walk through a real-world example. Keep it visual. Better yet, show them how delays or missing info in one stage affect the rest. People don't change behaviour because of policy - they change when they understand the impact on others.

19. How long should a work order stay in "planning" status?

As short as possible, without skipping steps. If a job sits in planning for weeks, something's wrong - parts unavailable, unclear scope, or lack of planner time. Flag these during backlog reviews. A healthy planning stage is 1–5 days for most jobs. More than that, and you risk losing momentum.

20. What if the same person submits vague requests over and over again?

Don't get frustrated - get proactive. Reach out, offer to walk them through one example request, and explain what "good" looks like. Often, it's a knowledge gap, not laziness. You might also create a cheat sheet or training guide for requestors. If they improve, the whole system improves.

PLANNER HQ

Made in the USA
Monee, IL
04 November 2026

67385907R10066